Army Distance Learning and Personnel Readiness

John D. Winkler

Henry A. Leonard

Michael G. Shanley

Prepared for the United States Army

Arroyo Center

RAND

The research described in this report was sponsored by the United States Army under Contract No. DASW01-96-C-0004.

Library of Congress Cataloging-in-Publication Data

Winkler, John D., 1953–
Army distance learning and personnel readiness / John D. Winkler, Henry A. Leonard, Michael G. Shanley.
p. cm.
"MR-1389."
Includes bibliographical references.
ISBN 0-8330-3040-X
1. Military education—United States. 2. Distance education—United States. 3. United States. Army—Personnel management. 4. United States. Army—Operational readiness. I. Leonard, Henry A. II. Shanley, Michael G., 1947–.

U408.3 .W5597 2001
355'.0071'5—dc21

2001048136

RAND is a nonprofit institution that helps improve policy and decisionmaking through research and analysis. RAND® is a registered trademark. RAND's publications do not necessarily reflect the opinions or policies of its research sponsors.

Published 2001 by RAND
1700 Main Street, P.O. Box 2138, Santa Monica, CA 90407-2138
1200 South Hayes Street, Arlington, VA 22202-5050
201 North Craig Street, Suite 102, Pittsburgh, PA 15213
RAND URL: http://www.rand.org/
To order RAND documents or to obtain additional information, contact Distribution Services: Telephone: (310) 451-7002;
Fax: (310) 451-6915; Email: order@rand.org

PREFACE

At the request of the Deputy Chief of Staff for Personnel (DCSPER), RAND conducted a project entitled "Personnel Policy Implications of Army Distance Learning." The goal of this effort was to help the Army maintain readiness and manage personnel efficiently as it implements distance learning (DL) throughout its Active and Reserve Component training systems. Initially, we described the potential scope of impact of the Army's distance learning program and established a framework for assessing personnel policy issues; that analysis documented the need to examine more specifically the potential of DL to improve readiness. Subsequently, we focused on ways to capitalize on DL to enhance personnel readiness, focusing on two areas: (1) DL's potential for reducing personnel shortages in Army enlisted occupations and (2) DL's effect on stability and the professional development of soldiers.

In this report we provide a broad overview of the research conducted over the course of this project and then discuss some ways in which the Army can further leverage the potential of DL technologies.

The research was conducted in RAND Arroyo Center's Manpower and Training Program. The Arroyo Center is a federally funded research and development center sponsored by the United States Army.

For more information on RAND Arroyo Center, contact the Director of Operations (telephone 310-393-0411, extension 6500; FAX 310-451-6952; e-mail donnab@rand.org), or visit the Arroyo Center's Web site at *http://www.rand.org/organization/ard/*.

CONTENTS

FIGURES

TABLES

SUMMARY

The Army is in the process of implementing TADLP, The Army Distance Learning Program, with an overall aim to improve the effectiveness and efficiency of military education and training. Distance learning (DL) uses technology to train soldiers where they are stationed, enabling the Army to deliver training more readily and also shorten the amount of time devoted to resident training at its schools. These changes are substantial and can have wide-ranging effects on how the Army trains and develops its leaders.

The Army is making major investments in distance learning technologies, building "distance learning classrooms" and developing courseware. These investments amount to approximately $850 million through 2015, including costs already incurred in both Active and Reserve Component programs. Observers have questioned the scope of these investments and whether the changes needed to manage such training are feasible to implement. Others have questioned more broadly the approach of TADLP: Is the program currently postured in a way to deliver on its promises?

This research, conducted for the Army's Deputy Chief of Staff for Personnel, examines distance learning from a personnel readiness perspective. The research team has explored aspects of personnel readiness that could be improved through distance learning, seeking ways to configure the distance learning program to realize those improvements. We examined two possible ways to leverage DL's potential to improve personnel readiness: (1) alleviate shortages of qualified enlisted personnel in critical military occupational specialties (MOSs) and (2) enhance the stability and professional development of soldiers and leaders. Our findings indicate that the Army should emphasize the exploitation of DL's capabilities in courses that promise high payoff in these areas. These include MOS skill-producing courses in specialties with chronic or large shortages, and officer and NCO development courses, especially those with large student loads and/or long residence requirements.

USING DL TO HELP REDRESS SHORTAGES OF QUALIFIED PERSONNEL

Active Component enlisted shortages pose a serious problem. For example, we estimate the Army was short more than 19,000 soldiers in operating strength in various military occupations in FY99; this was about 5.4 percent of authorizations. Another 8,500 were in jobs for which they lacked the full training specified for their skill and grade level.

Historically, the Army has addressed such problems by using accession, retention, and training strategies. In this report we shall focus on how DL might improve the effectiveness of training strategies aimed at reducing shortages. Such strategies include reclassification training at reenlistment, cross-training of soldiers, consolidation of military occupational specialties, and acceleration of training (especially for NCOs).

Reclassification is the transfer of soldiers from one MOS to another, with necessary training provided. In the context of reducing shortages, reclassification means transferring soldiers from surplus to shortage MOSs. RAND's analysis suggests that distance learning could assist in the reclassification process in several ways. First, DL could stimulate an expansion of the reclassification program, with a corresponding reduction in the number of MOS shortages. The reason is that reclassification courses taught using distance learning would be more attractive to (and therefore more utilized by) soldiers and commanders because they can take less time to complete and involve less time away from home. Second, reutilizing existing manpower to fill shortages increases the efficiency of the existing end strength. One way to value that gain is by the pay and allowances of soldiers moved from surplus to shortage positions—$32,000 per soldier per year for an E4 with 3–4 years of experience. In effect, this amounts to a better utilization of about $32,000 annually per reclassified soldier, since the soldier's time is now being employed in a shortage skill rather than in a skill where he was excess to the Army's needs. Hence, for every 31 soldiers so reclassified, the Army uses about $1 million in personnel costs in a way that more directly enhances personnel readiness. Third, the analysis suggests that DL could reduce the costs of reclassification by 30 percent. Cutting a 10-week course to 7 weeks produces an estimated cost avoidance of $4,500 per trainee. Finally, reclassification through DL can, in the long run, lead to the reallocation of some selective reenlistment bonuses (SRBs) to make them more effective in reducing shortages.

Distance learning can also improve the cross-training and consolidation processes. Both processes attempt to produce a more versatile soldier, one who can work effectively across a broader range of skills. While these processes do not necessarily decrease the number of MOS shortages per se, they do render shortages less damaging to personnel readiness. DL's effects in this area are much the same as they are in reclassification: reducing transaction costs (by offering shorter courses closer to home) allows for expansion of the program. As with reclassification, it also reduces training costs. DL also facilitates cross-training by enabling soldiers to access short refresher courses or on-line help when they encounter a less familiar aspect of their work. Such a resource can also help reduce resident time requirements in skill training programs.

DL can also make some parts of leader development training easier to accomplish in a more timely fashion; accelerated training, in turn, enables a reduction in the number of untrained personnel. As pointed out above, we estimate that in 1999, 8,500 soldiers held jobs for which they lacked full training, specifically the proper basic or advanced noncommissioned officer course. Distance learning can help by allowing the training to begin earlier in the select-train-promote cycle than the resident courses. This way, some portion of the work can be done before a resident course is available, and the course can be taken in small pieces at the home station. In many cases, the student will also be able to use the DL technology to test out of portions of the course, eliminating wasted time studying material already mastered. Moreover, DL can enhance self-development to accelerate the institutional training process.

ENHANCING STABILITY AND PROFESSIONAL DEVELOPMENT

Personnel turbulence—the disruption of stability caused by too frequent changeovers of positions—is a chronic impediment to readiness. DL-supported professional development courses can reduce turbulence and thereby promote stability. This effect can take one of two forms: either a modest increase in elapsed time between transfers (less time needed for schooling en route to new assignment), or a decrease in time spent away from duty position (in cases where the schooling is done on a TDY-and-return basis). Many consider the reduced time away from home and the lower travel costs as ancillary benefits of distance learning. Our analysis suggests these benefits are large enough to warrant greater emphasis on developing DL segments for Active Component officer and NCO professional development courses.

This portion of the research examined the institutional training and associated travel patterns of an officer's career course (Armor branch) to ascertain whether DL could offer stability and professional development benefits. Researchers considered four options, all of which were varying combinations of the two possible attendance modes: permanent change of station or temporary duty. Results show that using DL increases the time a student can be available to his or her home unit by an estimated 32 to 43 days. Some of the time at the home station would have to be devoted to DL study; allowing for this still leaves a net gain of as much as 15 to 27 days per student. Using conservative assumptions that all DL work would have to be accomplished on duty days and that the student would not be available for any duties on any of the DL days, we also developed a "worst case" estimate for days gained. This minimum estimate ranged from 8 to 15 days, depending on the option chosen. Extending this analysis to the entire population of captains who attend such courses each year yields between 115 and 135 additional working man-years available to the force, within the same total level of manpower. Extension to other officer and NCO courses would yield approximately 800 more man-years; together, these figures amount to well over 900 man-years. This is close to 5 percent of the student man-years we estimate the Army devotes to these kinds of courses each year.

Our analysis shows that travel and per diem cost savings will accrue only in those cases where DL can help shorten courses that are already being done in a TDY mode. These kinds of TDY savings could be in the $20 million range once the DL program matures. Cost savings may also be achievable in other aspects of institutional training, e.g., in the schools and the resources that support them. But too little is known about the tradeoffs in these areas to justify banking on such savings at this point.

FURTHER EXTENSIONS OF DL'S POTENTIAL

Our research provided additional insights into DL's potential. Using DL, the Army can help its people keep pace with the rapid changes that characterize today's world. Training on demand, provided through Web-based packages, will make it much easier for all the Army's soldiers to stay current in their fields, to refresh skills as needed, and to interact in near–real time with training institutions and with others in their fields, sharing ideas and techniques. Similarly, DL training packages can facilitate development of supplemental skills for soldiers assigned "additional duties." Also, more extensive use of DL to support military training opens more possibilities for using the same facilities, networks, and equipment to access civilian education courses as well.

POLICY IMPLICATIONS

Looking at distance learning from the standpoint of personnel readiness, we find that DL conveys potential benefits based on its potential to deliver self-paced training at a time and place best suited to the student. DL can provide needed flexibility to the training process, enable some training to be accomplished more quickly, and reduce the time needed for resident training.

Based on these findings, the Army may want to review the DL program and adjust implementation to capture more quickly some specific personnel readiness payoffs. First, we suggest more rapid adaptation of DL in enlisted occupations experiencing personnel shortages and for officer and NCO education. Moreover, we hold that the key advantage of DL is the significant additional flexibility it provides in the delivery of training at a time, place, and pace best suited to the individual student. The best way to leverage this capability is to use asynchronous[1] modules and make them readily available on Web sites. The Army also plans to provide synchronous training through DL, but this requires significantly more capital investment (i.e., for classroom facilities and hardware). We recognize that the Army has reservations about Web-based DL, and synchronous DL is likely to be the appropriate instructional mode in some instances. Moreover, the value of DL in these instances depends on whether the current quality of instruction and levels of training effectiveness can be maintained; this issue warrants further scrutiny. For example, the correct balance between distance learning and resident learning needs to be established and periodically checked for each course. Despite these concerns, we recommend that the Army carefully review the relative emphasis and resource allocations placed on these applications of distance learning.

---

[1]Synchronous DL has multiple students participating, usually with an instructor, at the same time. Seminars and group exercises are examples. Asynchronous training involves students individually; they participate at times best suited to them.

ACKNOWLEDGMENTS

We are indebted to many officials from Headquarters, Department of the Army, from TRADOC and its installations, and from other supporting agencies. In particular we would like to thank Maria Winston, Jim Coats, LTC Norvel Dillard, and the rest of the staff of the Training Requirements Division in the DCSPER's Directorate of Military Personnel Management. LTC Eli Alford and MAJ Gene Piskator of the Army's Personnel Command provided a wealth of useful data and helpful advice. We were greatly aided in our understanding of Armor Branch personnel structure and the Armor Center's distance learning initiatives by COL Richard Geier, Connie Wardell, Bob Bauer, Aubrey Henley, Leonard Fizer, CPT Chad Jones, and George Paschetto. Similar valuable assistance came from the Army's Signal Center's COL Larry Turgeon, SGM Dale Manion, Rosemary Berlin, Lilla Dancy, Howard Moore, Renee Carmichael, and Kelly Larsen. LTC Joe Charsagua of TRADOC was invaluable in helping us sift through the wealth of data we had been given by TRADOC and the Army in support of this project. Angela Lewis and John Bryant of the Program Management Office for TADLP provided helpful insights into the complex processes of cost and savings estimates.

We are also indebted to colleagues at RAND for helpful assistance throughout this research. Emile Ettedgui and Anders Hove provided important contributions throughout the course of the research. Rodger Madison provided helpful programming support, and Paul Steinberg and Jerry Sollinger improved the design and contents of briefings and documents. Finally, Herb Shukiar and Charles Kaylor offered helpful reviews of earlier drafts of this report. The authors alone are responsible for any errors or omissions.

ABBREVIATIONS

ACC	Army Competitive Category
ADS	Army Development System
AIT	Advanced Individual Training
ANCOC	Advanced Noncommissioned Officer Course
ATRRS	Army Training Requirements and Resources System
BNCOC	Basic Noncommissioned Officer Course
CAS3	Combined Arms and Service Support School
CD	Compact Disc
CSA	Chief of Staff of the Army
DCSPER	Deputy Chief of Staff for Personnel
DL	Distance Learning
HQDA	Headquarters, Department of the Army
HR	Human Resources
IET	Initial Entry Training
IPM	Inventory Projection Model
IPR	In-Process Review
MOS	Military Occupational Specialty
NCO	Noncommissioned Officer

ODCSPER	Office of the Deputy Chief of Staff for Personnel
OJT	On-the-Job Training
PCS	Permanent Change of Station
RL	Resident Learning
SL	Skill Level
SRB	Selective Reenlistment Bonus
TADLP	The Army Distance Learning Program
TCP	Transformation Campaign Plan
TDY	Temporary Duty
TRADOC	U.S. Army Training and Doctrine Command
VTT	Video Teletraining

Chapter One

INTRODUCTION

BACKGROUND

The Army has established The Army Distance Learning Program (TADLP) under the auspices of the U.S. Army Training and Doctrine Command (TRADOC). This program constitutes a significant investment in both distance learning (DL) courseware and infrastructure, amounting to about $850 million over the planning time span to 2015. This includes design and development of DL courseware for about 500 courses, along with an investment in infrastructure that includes networks and hardware for DL sites.

The Army believes there are a number of benefits associated with transforming its institutional training practices to incorporate DL. First, DL can reduce travel costs by moving training from the traditional schoolhouse to locations nearer to soldiers' duty stations. Second, by capitalizing on emerging educational technology and media like video teletraining (VTT), the Internet, and computer-aided instruction to deliver training, DL can offer needed training resources at multiple locations (rather than depending on a centralized schoolhouse). Third, DL enables more flexibility and continuity in the timing of training, by not requiring soldiers to leave their units for resident learning (RL) courses elsewhere. Finally, by moving training where the soldiers are, DL involves soldiers and their chain of command more fully in the training process. The chain of command will need to be involved to ensure that soldiers have time to complete their DL requirements. It may also be possible with some courses for the chain of command to influence the process of tailoring the course to soldier and unit needs.

Although new DL learning technologies promise to increase the amount, effectiveness, and efficiency of Army training, DL's ultimate success rests on the impact it can have in critical areas outside of training per se. One of these areas is personnel readiness. Always a key concern for the Army, personnel readiness is an important component of the Army's transformation goals. In particular, the first "axis" of the Army's Transformation Campaign Plan (TCP)—the mechanism for synchronizing the transformation vision within the Army—is "Trained and Ready"; it includes three "lines of operation" related to personnel readiness: Manning the Force and Investing in Quality People (number 3); Maintain Unit Readiness and Training (number 4); and Training and Leader Development (number 5). The document goes further in making the Deputy Chief of Staff for Personnel (DCSPER) the staff proponent for number 3, tasking that office, among other things, to

> Develop and implement a Personnel Transformation Strategy and Plan to ensure Army personnel support sustains and improves the Human Resource (HR) life cycle processes and HR information support systems for Army soldiers and civilians across the spectrum of Army operations.
>
> Determine means to attain and maintain required levels of personnel fill and stabilization throughout transformation.
>
> Execute approved personnel priority and stabilization measures.

DL should be able to help the Army—and the DCSPER as the personnel readiness staff proponent—to accomplish these tasks, in particular the task of "determin[ing] means to attain and maintain required levels of personnel fill and stabilization."

However, when we examined DL implementation within the Army, we found that the choice of DL courses in TADLP has not always been optimal from a personnel readiness perspective. This is particularly true when we look at the area of personnel fill. The Office of the Deputy Chief of Staff for Personnel (ODCSPER) maintains a list of critical military occupational specialties (MOSs), where shortages in required personnel call for close monitoring and special efforts to recruit, train, and retain more personnel. Table 1.1, drawn from the Arroyo Center's initial examination of the development plan for TADLP, compared the course list for TADLP against the ODCSPER's list of critical MOSs to see if TADLP had planned to offer DL courses for reclassification, Basic NCO Courses (BNCOCs), or Advanced NCO Courses (ANCOCs). The table compared "critical MOSs" identified during calendar year 1998 against the TADLP courseware development plan as of February 1999. At that time, planned courseware offered only modest help for training critical MOSs—only 19 out of 51 possible courses were planned for development.

We saw similar results when we looked at other measures of personnel fill, such as MOSs with the largest gaps between soldiers assigned and authorizations and ones suffering from low retention. In such

Table 1.1

DL Offered Modest Help for Training Critical MOSs (Monthly Readiness Reports, Calendar Year 1998)

MOS	Title	Was Courseware Planned?		
		Reclass	BNCOC	ANCOC
12B	Combat Engineer	No	No	No
14J	ADA C4I Tactical Operator	NA	No	No
14R	Stinger Crewman	**Yes**	No	No
31F	Electronic Switching System Operator	No	No	NA
31R	Multichannel Transportation System Operator	No	No	NA
31S	SATCOM Systems Operator	NA	No	No
45E	Armament Repairer	No	No	NA
45K	M1A1 Tank Turret Mechanic	No	No	No
45T	BFV Systems Turret Mechanic	No	No	NA
54B	Chemical Operations Specialist	**Yes**	**Yes**	**Yes**
63E	M1 Abrams Tank Systems Mechanic	No	No	**Yes**
63T	M2/3 BFV Systems Mechanic	**Yes**	**Yes**	**Yes**
67T	UH-60 Helicopter Repairer	**Yes**	**Yes**	**Yes**
77F	Petroleum Supply Specialist	**Yes**	No	No
92A	Automated Logistical Specialist	**Yes**	**Yes**	**Yes**
92Y	Unit Supply Specialist	**Yes**	**Yes**	**Yes**
98C	Air Traffic Control Operator	**Yes**	No	No
96B	Intelligence Analyst	No	No	No
98G	Voice Interceptor	No	No	No

SOURCE: HQDA, DCSPER, DAPE-MPE-DR.

situations, one might imagine that special efforts might be undertaken to fill these voids, perhaps by offering additional tools to train personnel needed at entry and advanced levels. At the time of our initial review, policies guiding the development of DL courseware did not emphasize applications like offering DL courses to support reclassification training of Active Component soldiers and NCOs. Our analysis did raise the question of whether TADLP *should* target such objectives, given their overall importance to personnel readiness.

OBJECTIVE

The research summarized here—and described in greater depth in two companion documents[1]—looks at DL's potential impact on two aspects of personnel readiness. One aspect, as suggested above, is whether DL can help the Army redress some of its personnel shortages. Also, the research considered another personnel readiness goal that derives from its inherent flexibility and potential advantages—whether DL can enhance stability and professional development for the Army's soldiers and leaders. Officer and NCO leader development courses are already a key focus of TADLP, with particular attention to Reserve Component personnel. The issue here is whether stability enhancement should be pursued as a major personnel readiness objective of DL, focusing particularly on Active Component officers and NCOs—if so, how might it be implemented, and what benefits might be accrued?

ORGANIZATION OF THIS DOCUMENT

In the companion documents discussed above, RAND researchers analyzed these aspects of personnel readiness in some detail.[2] We found that

- Distance learning technologies can complement and improve three of the Army's strategies for reducing enlisted personnel shortages (providing reclassification training, supporting cross-

[1]See Leonard et al. (2001) and Shanley, Leonard, and Winkler (2001).

[2]See the companion documents for a complete discussion of the methods used in the evaluation.

training and MOS consolidation, and accelerating training to Active Component soldiers and NCOs);

- Converting portions of the Army's resident courses to DL can enhance the stability of soldiers by increasing time spent at home station, reducing time spent in training institutions, and potentially saving travel costs associated with resident training.

These two key findings are discussed in more detail in Chapters Two and Three. Chapter Four takes a broader look at how DL can help the Army improve personnel readiness. Chapter Five offers conclusions and some cautionary insights.

Chapter Two

DL CAN ENHANCE EXISTING STRATEGIES FOR ALLEVIATING PERSONNEL SHORTAGES

Enlisted personnel shortages in the Active Component operating strength are a significant readiness issue documented in a number of reports and Army personnel files, including the Chief of Staff of the Army's (CSA's) monthly readiness reviews. Shortages are best defined and measured statistically as the gap between the number of assigned soldiers in a particular grade and skill and the number of positions authorized to have a soldier of that grade and skill assigned. There are several ways of aggregating this measure, but all of them indicate that the overall number of shortages in the Army is sizable.

First, at the military occupational specialty (MOS) level of detail, we estimated that the Army was short about 19,300 soldiers in FY99, representing 5.4 percent of all authorized positions in the enlisted force. About half these shortages can be traced to a shortfall in total Army personnel relative to authorizations for a given MOS; the other half is created by an imbalance among occupations: overassignment in some occupations and underassignment in others. A second kind of shortage occurs when fewer NCOs are available than are needed to fill authorized positions of a given grade. In this case, positions may be filled by personnel of a lower grade who have not received the education required for the higher-grade position. We estimated that about 8,500 E6 and E7 positions, representing about 2.5 percent of all authorizations, were occupied by NCOs who had not yet received the NCO course required for those jobs.

Shortages have a number of causes, including recruiting shortfalls and low retention rates; an increase in requirements in a particular skill can have the same effect. The training process can also contribute to shortages, as when attrition from MOS-qualifying courses is high or when a shortage of training equipment or instructors keeps the Army from training as many qualified candidates as it would like. This can also happen when time required for training increases—a not unusual result of increased sophistication. The overall result in any of these cases—shortfall of skilled soldiers—is essentially the same, as is the potential for DL to help address the problem.

The fact that there is an ongoing problem with shortage MOSs is not news to the Army. The Army has several strategies to deal with shortages. Two strategies—increasing accessions and increasing retention—address the problem by getting more soldiers into the pipeline and keeping more soldiers from dropping out during or at the end of their enlistment contracts. These strategies employ targeted recruiting and retention efforts, frequently with enlistment or reenlistment bonuses. Additional strategies are aimed at altering training and occupational structure. These include expanding the number of training seats during initial entry training (IET), offering reclassification training after IET, providing cross-training, consolidating MOSs to yield fewer positions, and accelerating the pace of education and training (specifically, offering NCO courses sooner to soldiers at advanced grade levels).

Given these strategies, how could DL be used to help the Army alleviate MOS shortages? We argue that *using DL to provide reclassification training, support cross-training and MOS consolidation, and accelerate education of NCOs offers a great deal of potential.* In this chapter we support this argument by describing the potential effects of DL-based strategies in terms of two measures of effectiveness: fill rates and costs. We concentrated our analyses on some specific examples: UH-60 Helicopter Repairer (MOS 67T), and several MOSs in the Signal area. These analyses were supported by individual inventory projection models (IPMs) to measure the long-term impact of DL-based strategies on the size of the MOS inventory.[1]

---

[1]For more details on the IPM approach used, see Shanley, Leonard, and Winkler (2001).

APPROACH

Our analytical approach started by identifying three strategies the Army currently uses to address personnel shortages and that DL could potentially improve: (1) reclassification, (2) cross-training/ MOS consolidation, and (3) acceleration of training. We then chose several cases for more detailed examination in relation to the strategies: MOS 67T (UH-60 Helicopter Repairer), and three MOSs in the Signal area, 31F (Electronic Switching System Operator), 31P (Microwave Systems Operator/Maintainer), and 31U (Signal Support Systems Specialist). Once we completed the analysis of the sample MOSs, we turned our attention to estimating the potential forcewide effects of using DL to address personnel shortage issues.

In assessing how DL-based training strategies might affect shortages, we chose two primary measures of effectiveness (MOEs): changes in shortages or fill rates, and changes in costs per shortage filled. To support our estimate of the change in shortages, we developed IPMs for the occupations under study and for the force as a whole. To support our estimate of the change in costs, we developed a suitable methodology based on early military experience with DL courses.

A note of caution is appropriate here: our analyses include an assumption that DL's potential can be fully realized without causing any reduction in the quality of training. We also note that past research supports the contention that DL, when properly implemented, can provide training as effectively as the classroom training it replaces.[2] However, there can be no doubt that DL's introduction will bring about large and fundamental changes (technically, organizationally, and culturally) in how training is conducted in the Army. This will be true particularly, but not exclusively, for institutional training. Viewed in this way, DL clearly poses some risks to the quality of training, especially during the transition period. The keys to maintaining training quality and achieving desired learning out-

---

[2]A considerable amount of literature exists to substantiate that DL, when judiciously applied to teach skills suited to such training, can produce the same level of training effectiveness, and can even accomplish this task in less time. See, for example, Keene and Cary (1992, pp. 97–103), Farris et al. (1993), and Phelps et al. (1992, pp. 113–125). Orlansky and String (1979) also provides an in-depth treatment of this and related topics. More condensed discussions can be found in Orlansky and String (1981) and Orlansky (1983).

comes under DL will continue to be careful implementation and monitoring, as well as sustained emphasis and support throughout the Army.

RESULTS

How DL Can Improve the Effectiveness of Reclassification Training

Reclassification of serving soldiers (i.e., the transfer of soldiers from one MOS to another, with necessary retraining provided) is an important means of addressing shortages in certain skills. Historically, the Army reclassifies a sizable number of soldiers each year. In FY99, the total number of reclassifications was 5,220 soldiers (about 7 percent of the size of total accessions). In addition, an estimated 2,910 prior-service accessions were trained to fill MOS shortages; this is, in essence, another form of reclassification.

The Army uses the reclassification strategy because it has a number of perceived benefits. First, in reducing shortages, it is more productive for the Army to reclassify a soldier than to bring in a new soldier through the accession process. A key reason is that reclassification training involves many fewer training losses than the accession process. Second, reclassification can be targeted to reduce shortages for Skill Level 1 (SL1) and NCOs alike. Third, reclassification is more efficient than accession, primarily because it avoids some accession costs (e.g., enlistment bonuses or other incentives) and the expenses of basic training.

What is the potential value of DL reclassification? First, our analysis suggests that DL could stimulate an expansion of the Army's reclassification program, with a corresponding reduction in the number of MOS shortages. The basis for this expectation boils down to a transaction cost argument: because DL lowers the transaction costs both for potential trainees and for the organizations that must bear the costs, more will choose the reclassification alternative. To see how transaction costs are lower, first consider potential trainees. Obstacles to entering a new occupation are reduced because DL courses take less time to complete and involve less time away from home and family than the traditional advanced individual training (AIT) alternative. Table 2.1 shows the example of MOS 67T. The DL-supported course is 43 percent shorter in total (8 weeks, three days versus 15 weeks) than the corresponding AIT course. Moreover, the 67T course is 72 percent less in terms of time away from home—only 4 weeks, one day.[3]

Second, consider the transaction costs for the Army's units and organizations. These units, like the individual trainees, gain from a course that takes less time to complete and from a training scenario that allows some access to the soldier during the training period. Moreover, units (and the Army as a whole) are also better off because DL reclassification training costs less. Figure 2.1 continues with the example of 67T, comparing the cost of accession, the cost of AIT reclassification, and the cost of DL reclassification. AIT reclassification costs are 36 percent less than accession costs for 67T because (as argued above) the reclassification option avoids large enlistment bonuses and the cost of basic training. In addition, AIT reclassification saves on student pay costs during training (i.e., the pay and

Table 2.1

DL (TADLP) Versus AIT Course Characteristics: The Example of the 67T Reclassification Course

Characteristic	AIT Course	DL Course (TADLP)
Total course length	15 weeks	8 weeks, 3 days
Residential length	15 weeks	4 weeks, 1 day
DL length	None	4 weeks, 2 days
Testing out of already-mastered material	No	Potentially yes
Potential obstacles	Funding Training seats Equipment	Cost of added course development

SOURCE: TADLP In-Process Review (IPR) November 1999.

[3]In making these comparisons, we implicitly assume that the DL reclassification course is as effective as the AIT course in giving established soldiers (who are changing their occupation) the required skills for the 67T MOS. Recall our previous citations regarding the comparative effectiveness of DL, and the importance of making the right choices as to which skills can be taught using DL.

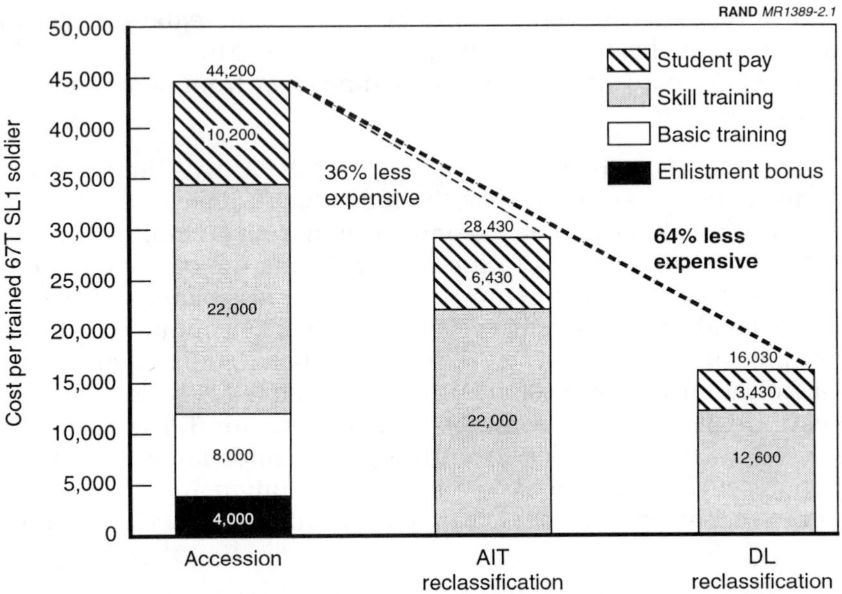

Figure 2.1—DL Reclassification Is the Least Expensive Way to Alleviate SL1 Shortages

allowances of trainees in the process of becoming SL1-qualified).[4] Moreover, DL reclassification training is more than 40 percent less expensive than AIT reclassification training (and 64 percent less expensive than accession—see the third bar in Figure 2.1)—because (as shown in Table 2.1, above) the class is substantially shorter.

While the forcewide benefits of expanding reclassification depend on how large the program can ultimately become, our analysis suggests that forcewide gains would be substantial. First, enhancing the Army's ability to reutilize existing manpower to meet force structure requirements increases the efficiency of the existing end strength. One way to value that gain is by the pay and allowances of soldiers moved from surplus to shortage positions: $32,000 per soldier per year for an E4 with 3–4 years of experience. In effect, this amounts to

[4]Those costs are directly proportional to the length of training in each case and thus are lower in the reclassification case, where IET time is avoided.

a better utilization of about $32,000 annually per reclassified soldier, since the soldier's time is now being employed in a shortage skill rather than in a skill where he was excess to the Army's needs. Hence, for every 31 soldiers so reclassified, the Army uses about $1 million in personnel costs in a way that more directly enhances personnel readiness.

Second, our analysis suggests that DL could reduce the training cost of reclassification by about 30 percent for the average course,[5] amounting to significant savings. Assuming an average 10-week AIT course reduced to a 7-week DL reclassification course, we estimate that the training costs (in terms of military personnel and expenditures for operations and maintenance) avoided in filling shortages would amount to $4,500 per additional soldier reclassified, or a million-dollar savings for each additional 222 reclassifications. Moreover, if current reclassifications (and prior-service accessions) could eventually use a DL training mode, we estimate (using the same $4,500 per soldier figure) that savings compared to the current AIT alternative would amount to $29 million per year.

An indirect effect of an expanded reclassification program could lead to a third forcewide benefit. Specifically, we find that expanding reclassification through DL could lead to a longer-term opportunity to reallocate some Selective Reenlistment Bonuses and thereby increase their effectiveness in reducing shortages. We estimate that 5 percent of the 13,500 soldiers receiving bonuses in FY99 were NCOs in occupations with shortages at the SL1 level, but not the NCO level. In such cases, the eventual movement of DL-stimulated SL1 reclassifications to NCO positions could lead to a surplus of NCOs in that occupation. The surplus, in turn, would allow reallocation of SRBs to effect reduction in shortages in other occupations. Assuming the affected NCOs received the average bonus amount, $6,700, the potential for SRB reallocation due to DL could amount to as much as $4.5 million per year.

---

[5]This assumes that, on average, RL and DL training cost the same on a daily basis, but that DL courses can be 30 percent shorter than RL courses with no degradation in training quality. Research previously cited (see in particular Orlansky and String (1979, 1981)) lends credence to the 30 percent efficiency gain, again with the caveat that this applies only to those elements of a course suitable for DL. We also examined some of the early course designs in TADLP, and found an average reduction of close to 30 percent. Shanley, Leonard, and Winkler (2001), Appendix B.

How DL Can Improve the Effectiveness of Cross-Training and MOS Consolidation

Cross-training and MOS consolidation are both attempts to produce more effective soldiers, capable of performing a broader range of activities. With cross-training, soldiers already proficient in one MOS are trained to perform related activities in another MOS so that they can informally fill in for that other occupation when necessary. With MOS consolidation, MOSs that perform similar activities are formally combined into one occupation, and soldiers in each of the old MOSs are given additional training to become proficient in the required skills of the new MOS.

Although not employed extensively in the recent past, there is evidence—both from a previous RAND study and from the civilian sector—that cross-training and MOS consolidation represent a feasible strategy when properly implemented. Moreover, they can also be cost-effective strategies for alleviating the effects of personnel shortages. They accomplish this by increasing the skill base of soldiers to make both soldiers and the assignment process more effective and efficient. In the case of consolidation, reducing the number of MOSs simplifies the assignment process, reducing force structure imbalances and allowing a direct reduction in shortages. In the case of cross-training, an increased amount minimizes the impact of personnel shortages by helping redress imbalances in workload. Thus, while the actual number of MOS shortages may not decrease as a direct result of cross-training, these shortages are rendered less damaging to personnel readiness.

What is the potential value of using DL to deliver the additional training required for cross-training and MOS consolidation? The effects are similar to those DL has on the reclassification strategy. First, DL allows for an expansion of the strategy, accompanied by a corresponding reduction of the impact of shortages. The reason is that DL can reduce the transaction costs of training soldiers compared to resident learning (RL) by offering a shorter course closer to home; lower transaction costs will, in turn, increase the feasibility of using the strategy.

Second, DL could decrease the cost of cross-training and consolidation. In the case of cross-training, reductions in cost could be even greater than those from DL reclassification. Cross-training, by definition, is dealing with functionally similar MOSs, whereas reclassification might involve two entirely different MOSs. With more overlap between the new and old skills, cross-training could make efficient use of the modular aspect of DL, allowing soldiers to avoid the parts of the reclassification course that cover tasks they have already learned, reducing even further course length, training repetition, and temporary duty (TDY) time.

For MOS consolidation, the way DL would help with training costs depends on how the consolidation is accomplished. If two functionally similar MOSs are simply combined into one (perhaps because of technological change), the use of advanced learning technology might contribute to the development of a feasible training strategy by reducing the time required to complete it. But if the objective is to produce a generic specialist across two or more specialties (as is true, for example, for helicopter repairers in the civilian world), DL could provide much or all of the equipment-specific training that soldiers need for a specific assignment without taking them away from their home station.

From a forcewide perspective, the potentially large increase in MOS consolidation in the near future suggests that this strategy can have a large impact in avoiding future training costs. For example, the organization of future aviation brigades suggests compatibility with MOS cross-training and MOS consolidation. More broadly, the results of the ADS XXI Task Force indicate that the Army plans to reduce the number of MOSs from 241 to 189. Since only a small part of this reduction is a net of deletions (23) over additions (12), this change if approved and implemented will result in a significant consolidation of many MOSs.

How DL Can Improve the Effectiveness of Accelerating Training

When NCOs cannot be trained in a timely way, the result is a further shortage of trained personnel. For FY99, we estimated 8,500 E6 and E7 positions occupied by soldiers not formally trained (with the appropriate Basic NCO Course (BNCOC) or Advanced NCO Course (ANCOC)) for those jobs or not trained for their grade. That number represents 2.4 percent of all authorizations, and 8.9 percent of E6 and E7 authorizations. Some of these are NCOs not yet formally trained for their grade, but most are "fast-trackers" serving above their grade in positions for which there would otherwise be a shortage.

With proper support and monitoring, DL could allow BNCOC and ANCOC training earlier in the select-train-promote sequence. First, DL training can begin before scheduled resident training courses are available. Second, DL training can be taken in small pieces, on a more or less continuous basis. Third, DL training can occur at home station. Fourth, modularized DL courses would allow "testing out" of already mastered material, which means fast-trackers who get much of their experience through on-the-job training (OJT) would not have to sit through the parts of course material they have already mastered. Finally, DL can enhance the type of self-development training that can accelerate the institutional training process.

Accelerating BNCOC and ANCOC can also have the indirect benefit of reducing some grade-specific occupational shortages. This point is illustrated in Figure 2.2, which shows how more timely ANCOC training for the Microwave Systems Operator/Maintainer (MOS 31P) could speed up full qualification of E6s serving in E7 positions, possibly enabling more rapid promotion as well. The figure also shows that all NCO gaps are not at the E5 level. For example, 31P has almost exactly the right number of E5s assigned, but a significant shortage of E7s. In this case, the use of reclassification or cross-training would not be a cost-effective way to address the shortage. But one might use a DL-based strategy of training and promoting E6s faster to fill the E7 shortage, especially if there are E6 fast-trackers already filling E7 positions. The training vehicle used would be a DL version of the existing 31P ANCOC course, which would presumably allow greater access to training materials, with a significant portion of the course to be taught at the student's home station. Upwardly substituted fast-trackers, who are already performing some of the skills taught in the course, could test out of some sections of the course or be allowed to continue to learn while on the job, with the full support of schoolhouse instructors at a distance.[6]

---

[6]The effect of accelerating training on grade-specific fill rates is limited by the size of the surpluses in the lower grades. For example, in Figure 2.2, the E6 surplus operating strength is 13 NCOs (over and above soldiers in the Trainees, Transients, Holdees, and

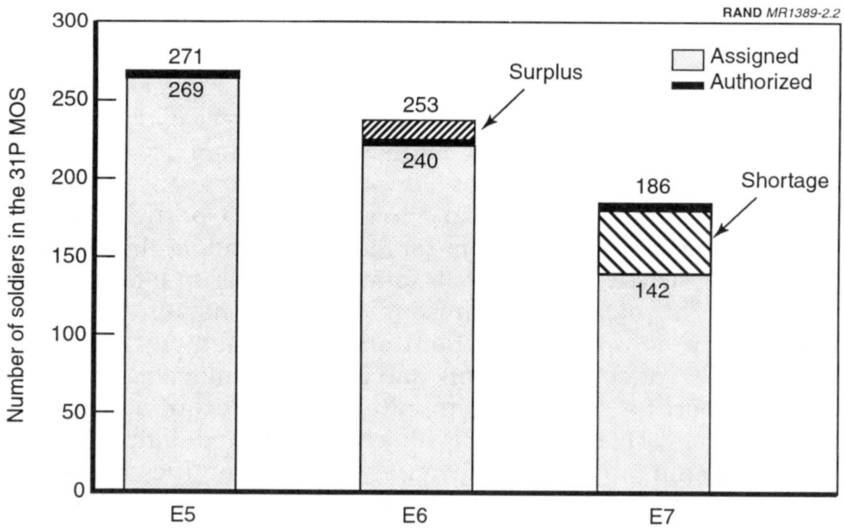

Figure 2.2—Using DL to Accelerate NCO Education Can Reduce Shortages at Grade-Level Bottlenecks

Given the extent of training shortages among NCOs, we believe DL has a relatively high potential to reduce shortages by accelerating training. Moreover, with the new NCO Education System (NCOES) model projecting more individual training for NCOs, we think there is additional potential for DL's support of BNCOC and ANCOC to increase in the future, helping to avert future training shortages.

Keys to Realizing DL's Potential in Reducing MOS Shortages

Using DL in the manners described above offers significant potential for further reducing shortages of qualified personnel and for reducing the marginal cost of achieving those reductions. Realizing this potential requires implementing the DL program in ways that are most likely to produce the available benefits. This means early selection of courses for conversion that will do the most to reduce the shortage problem (i.e., courses, especially longer courses,

Students (TTHS) account), while the E7 shortage is 44 NCOs. Thus, accelerated training could address up to 30 percent of the E7 shortage.

focused on shortage MOSs, consolidating MOSs, and MOSs with ANCOC and BNCOC backlogs). Most important, both now and in the longer term, it means creating DL courses with sufficient flexibility to be easily integrated into varying soldier career paths. In this regard, the DL program should emphasize maximum use of emerging learning technologies to help reduce learning time (thus shortening overall course time) and to allow significant portions of the training to be completed at home station. In addition, the DL program should strive to avoid pitfalls found in the past in industry and academia. This means providing sufficient student support to ensure speedy completion without increased personnel tempo (PERSTEMPO) or course attrition, and providing sufficient administrative support for scheduling, monitoring, and recording training results. Finally, DL needs to provide courses as modularized, "just-in-time" training to take full advantage of opportunities to reduce unnecessary training and to provide an extensive capability for refresher training on demand.

Chapter Three

DL CAN ENHANCE LEADER STABILITY AND AVAILABILITY AS PART OF ARMY PROFESSIONAL DEVELOPMENT COURSES

Just as shortages of qualified personnel are a readiness problem for the Army, so too is turbulence—a lack of stability—because the more frequent the turnover of unit personnel, the less chance there is to develop cohesion and teamwork, which most consider to be key elements of an effective unit.

Turbulence is a stubborn problem. First, achieving zero turbulence is inherently undesirable, since an Army with near-zero turbulence is a stagnant one with limited upward or lateral mobility and few opportunities for broadening or advancement. Second, many measures that might reduce turbulence to "better" levels would be distasteful or undesirable in other ways. These include increasing overseas tour lengths, altering the force structure, decreasing the frequency of promotions, reducing permanent change of station (PCS) school opportunities, and reducing individual taskings.[1]

However, DL-supported training programs—i.e., the judicious substitution of DL segments for some residential training—provide an option for enhancing stability without undesirable policy or structural changes. *Converting portions of the Army's resident courses to DL can enhance leader availability and stability.* While this overall result is more or less intuitive, our analysis illustrates the magnitude of that result. DL-supported courses increase soldier availability not only by reducing the time spent away from home station, but also by reducing the total amount of time needed for the training. And doing so also improves family quality of life. Moreover, reducing the length

[1]See Hix et al. (1998).

of time students are in TDY status for residential training will also produce some modest cost savings; the savings potential is less clear when we look at the possibility of converting courses from PCS to TDY.

In this chapter we show the results of an analysis that focused first on one officer course, the Armor Captains Career Course, to determine the potential effects of DL. Captains career courses consist of a branch-specific advanced course, taught at the branch school, and a course for junior staff officers called the Combined Arms and Service Support School (CAS3), taught at Fort Leavenworth. The Army already offers a Reserve Component Armor Officer Advanced Course that depends heavily on DL, with only a two-week resident phase. For the Active Component career course, we looked at the effects of converting 25 percent[2] of the resident portion of the advanced course segment to DL and applied our previously noted 30 percent time reduction factor to that portion of the course sequence. Applying these factors to the Armor Officer Advanced Course results in a 16-day DL phase and a 94-day resident phase.[3]

With this course structure as a basis, we explored four options to modify current practice. Today's pattern includes a resident advanced course of 18 weeks, 6 weeks at Fort Leavenworth for the CAS3 course, and, for most students, a three-week tank or cavalry troop commander's course. The alternatives explore different combinations of PCS and TDY status to determine a range of possible effects on time at home station (i.e., increasing stability) and some of the relevant costs.

---

[2]TRADOC currently envisions converting about 25 percent of officer advanced courses to DL mode. In addition, we note that the Armor School is conducting an advanced course for RC officers with upwards of 70 percent of the instruction done by DL.

[3]We considered DL conversion only for a portion of the 18-week advanced course phase: 18 weeks × 40 hours/week = 720 hours. A 25 percent reduction in this resident time yields a resident phase of 540 hours, 13.5 weeks, or about 94 days. The remaining 180 hours (22.5 days), when reduced by the 30 percent factor, come to 15.75 = 16 days for the DL phase. So the student will be at home station about 4.5 (18 – 13.5) weeks longer, and about 16 days of that time will need to be devoted to DL study. If partial conversion of the other phases also proves possible, the time savings will be greater.

Once we completed the analysis of the Armor Captains Career Course, we extended the results to other courses where they could apply.

DL CAN INCREASE TIME ON STATION FOR THE ARMOR OFFICER CAREER COURSE

As noted above, we considered four options to modify the current practice:

- Option 1: All resident requirements completed in TDY status;
- Option 2: Resident requirements done as three separate TDYs;
- Option 3: Mix of PCS and TDY-and-return, as determined based on Army requirements and officer preferences;
- Option 4: Current PCS pattern with career course segment shortened by including a DL module.

The motivation behind Option 1 was to start with a comparison of TDY versus PCS, assuming that the DL-induced shortening of the advanced course could make a conversion to TDY feasible. In this option, all the courses in the career course[4] sequence would be completed in a single (and lengthy) TDY. This case imposes significant family separation. Thus, Option 2 allows for return home between the different segments; this reduces the family separation impact somewhat by breaking up the separations. But it adds more travel costs, and the family separation is still a major drawback. Option 3 allows some additional flexibility between PCS and TDY, allowing the Army to help officers avoid family separations but also allowing the possibility for others to avoid family disruptions by taking the course TDY and moving their families once instead of twice.[5] Option 4 avoids altogether the family separation effects by simply leaving the current PCS pattern in effect and shortening the

---

[4]Note, from our earlier discussion, that the career course program includes CAS3, and that most Armor career course students also complete a tank commander's or troop commander's course while at Fort Knox.

[5]We expect most officers would want their families to accompany them. This option would make it possible for the Army to accommodate those with important reasons—like spouse employment or timing of schools for spouses or children—to do otherwise.

advanced course. This would be the easiest change to make administratively.

For each option, we determined how much total time at home station would increase (i.e., available days). However, since some of the time at home station must be devoted to the DL instruction, we calculated the amount of time DL studies would take and derived a second number, showing how much additional time would be available to units. Finally, we calculated a "worst case" number. This number is based on a strict assumption that the student would be unavailable to the unit during any of the time devoted to DL. In reality, it is quite likely that an officer could participate in some unit activities with no detriment to the DL study, even when concentrating on DL.

As shown in Figure 3.1, the four options increase the time on home station—compared to current practice—by between 32 and 43 days per officer, depending on the mix of PCS and TDY status, shown by the gray columns). Stripping out the time necessary for DL yields between 16 and 27 additional days available to the unit (shown by the black columns). The worst case is between 8 and 15 days (shown by the white columns), again depending on the PCS/TDY mix chosen.[6]

The effect on quality of life (narrowly defined here as time available at home with families) varies among options. If the entire course is done in a TDY status (Options 1 and 2), the officer spends more time away from his or her family than with the current course, which is done in a PCS mode. Option 3 causes no family separation for students who wish to move their families—it is similar in this regard to both current practice and Option 4—and permits other students to stabilize their families if desired.

In our sensitivity analysis, we analyzed the effects of changing the size of the DL module, and thus, as a derivative, the length of the residential portion and duration of the overall course. There is a

---

[6]Calculated by removing weekend time from the overall estimate of additional available days. For example, 43 additional days span 6 weeks, including 12 weekend days. If we assume no weekend work, saved days come down to 15 (43 – 12 – 16); similarly, 32 available days less 16 days for DL and 8 weekend days yields a minimum of 8 days saved. We believe this to be a very conservative assumption, but we use it as a basis for minimum estimates.

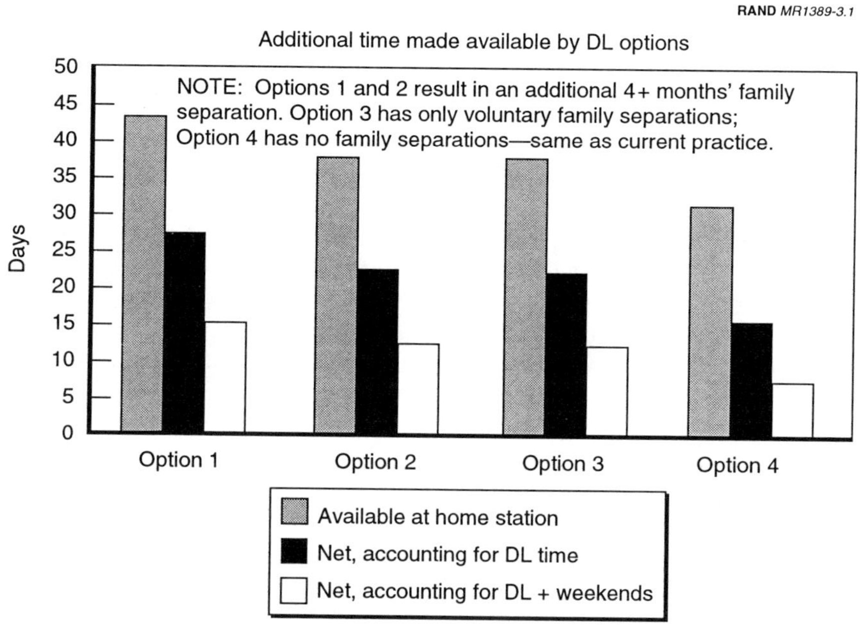

Figure 3.1—Each of the Options Yields More Available Days

linear relationship between the number of days an officer is available (by all three of our measures) and the size of the DL module: the larger the module, the more available days. While this suggests that more DL conversion is always better, this is of course not true. The limit to conversion will come from deciding how much can effectively be converted to DL out of a curriculum that includes a host of potentially complex topics. This is an area that the Army is already exploring more extensively; our analyses illustrate the potential benefits of further conversions to DL if they can be supported from the standpoint of training effectiveness.

COSTS USING DL ARE ROUGHLY COMPARABLE TO CURRENT PRACTICE

We also compared some of the relevant costs of the various options to those of current practice. We focused on savings that can reasonably be estimated based on envisioned DL conversions and their effects on time spent and travel involved. Key elements in our estimates included factors for PCS costs, TDY travel, lodging, and per diem. The other determinants were the number of PCS moves (for courses that involve PCS moves, like the advanced courses), and the duration of the TDY periods.

Considering these elements, the cost of the various options does not vary much from the current practice. Figure 3.2 illustrates this, showing that three of the four options considered are more expensive than the current one, but only slightly so. The primary difference is that these options involve significantly more TDY (the lower two bands in each column), and these costs more than offset the PCS savings.

In our sensitivity analysis, we looked at how sensitive the options were to changes in course length and PCS or TDY rates. TDY options

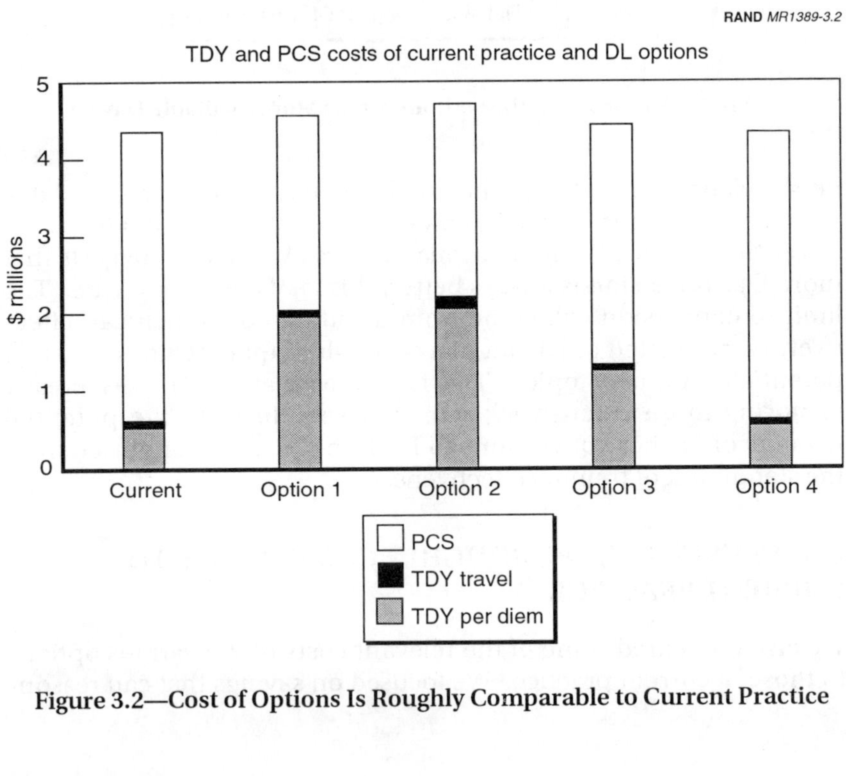

Figure 3.2—Cost of Options Is Roughly Comparable to Current Practice

naturally tend to be favored when PCS costs are higher and course length is shorter; likewise, PCS options cost less when courses are longer and TDY rates are higher. Our analysis showed that within reasonable ranges for the cost factors, the PCS options cost about the same as or less than the full-TDY options.

STABILITY ENHANCEMENT CAN BE SUBSTANTIAL ACROSS ALL CAREER COURSES

How do these increases in available days transfer across other career courses? Using the numbers for the fourth option and applying the estimated per-officer increases in available days to a population of some 3,500 to 4,100 captains per year going to career courses[7] yields an estimate of about 300 to 360 additional man-years (using the total time on station measure), or between 115 and 135 *working man-years*[8] (using our minimum measure, converted to working years). Consistent with our previous reasoning, these effects can be larger than our conservative measures indicate, depending on how carefully officers and their supervisors can schedule DL preparation and study time around unit duties. Also, the effect on stability enhancement is greater if TDY options are chosen (less time used in PCS processing), but the effect on families is ambiguous at best and more likely negative. Overall, costs for student travel will also rise in the TDY alternatives.

ARMOR CAREER COURSE FINDINGS EXTEND TO OTHER COURSES

Analysis to this point has focused only on courses now being done in PCS mode, showing that partial DL conversions of these courses enhance stability by keeping soldiers in their units longer at about the same costs (PCS or TDY) now being incurred. Extending the analysis to other courses done in a TDY mode only, we find the same

[7]The FY99 Army Competitive Category (ACC) promotion list had about 4,100 names on it; FY98 had about 3,500.

[8]We use a figure of 240 working days for a working year. If we are using a measure—like our lowest measure here—that discounts weekend days, then it is appropriate to use a working-year factor that similarly discounts nonworking days.

general implications for stability enhancement, and unambiguous savings estimates as well.

Table 3.1 illustrates this for two BNCOCs—Artillery Fire Specialist (13B) and Signal Support Systems Specialist (31U). The current 13B BNCOC requires 43 residential days, including weekends. Applying our course conversion methodology, we find this course would break into a residential module of about 99 hours and a DL module of about 104 hours. Total time required, including weekends, to complete the residential portion is about 17 days. The 104 DL hours require 13 8-hour days. Thus, the NCO will spend 26 (43 – 17) more days at home station and could be available to his unit for a maximum of 13 (26 – 13) days. Minimum additional days are estimated at 6, after allowing for the presence of weekends as previously discussed. When we convert these per-person figures to man-years, we use work-years for the last (lowest) measure to be consistent with our removal of weekend days. The TDY cost savings is simply the saved

Table 3.1

Analysis of Artillery and Signal BNCOC

	Artillery Fire Specialist (13B)	Signal Support Systems Specialist (31U)
FY99 attendance	153 TDY and return 17 TDY en route	111 TDY and return 18 TDY en route
Original course length	43 days	77 days
DL length	17 days residential 13 days DL 30 total days	30 days residential 23 days DL 53 total days
Increase in available days		
Max	26 per person (4,420 man-days/ 12.1 man-years)	47 per person (6,100 man-days/ 16.6 man-years)
Net	13 per person (2,210 man-days/ 6.0 man-years)	24 per person (3,100 man-days/ 8.5 man-years)
Min	6 per person (1,020 man-days/ 4.3 work-years)	11 per person (1,400 man-days/ 5.9 work-years)
Estimated savings (lodging and per diem)	$133K	$182K

residential days times the TDY cost factor ($30) times the student load, e.g., 26 × $30 × 170 = $132,600 or $133K. The process for the 31U course is analogous.

Other courses such as reclassification training or short-duration courses also show benefits. However, since benefits are generally proportionate to course length, shorter courses naturally yielded smaller benefits. Overall, we estimated on-station man-year increases to be just under 2,400, with an increase in available work-years (our minimum measure) of about 800.

Chapter Four

POTENTIAL BENEFITS OF DL EXTEND TO OTHER AREAS

DL's benefits are not limited to the potential uses described above. The principal advantage of distance learning—an advantage with far broader application than professional military training and education—is the ability to deliver training efficiently in a manner that can be tailored to the student's individual needs in terms of timing, content, focus, and pace. While this feature would be beneficial in virtually any education scenario, it is even more valuable in today's environment of rapid technological growth, frequent deployments, and other distractions of military personnel from assigned duties, as well as in an environment that requires leaders with broader knowledge and skills.

A common belief expressed in education literature is that continuous education is more valuable than education presented in packages at discrete intervals. The hypothetical ideal is one-on-one instruction presented by an expert tutor, as needed and when needed by the student. DL makes a near variant of this ideal feasible: packages a student can schedule when needed and when convenient, with content the student can pick from to focus on areas where improvement or greater familiarity is required, and with some real-time feedback. These general benefits of DL can obtain in virtually all applications.

MILITARY TRAINING BENEFITS

DL's ability to provide training "on demand" is one of its greatest potential contributions. Trainers and educators in and out of the military services have explored and begun to exploit the potential of paper-based and compact disc (CD)-based job aids. Web-based DL goes further, enabling training proponents to deliver training more responsively (instantaneously, for asynchronous training), update materials, and monitor their usage and usefulness. Job aids in any of these forms can be used to provide "just-in-time" skill training. In many cases, this is likely to be more useful than more formal residential training, which involves waiting for a scheduled opening and then a protracted absence from assigned duties. DL similarly makes refresher training more readily available, and this has important implications for an Army increasingly dependent on skills that can quickly become outdated because of technological developments.

The pace at which technology evolves makes it largely infeasible, if not completely impossible, to employ residential training to keep skills fully up to date in many technical specialties. In addition, military modernization and equipment procurement schedules can move so rapidly that soldiers and leaders find themselves needing to deal with a new generation of systems (or new systems altogether) every time they move from one location to the next—and frequently even before they move. While technical libraries, telephone assistance lines, and mobile training teams can help as needed to upgrade and adapt the skills of people in the field, DL can accomplish the same requirements—or a substantial portion of them—more rapidly and more efficiently.[1]

The military services are all to some degree developing training-support capabilities in which the proponent for each technical skill maintains a Web-based course covering all aspects of the skill that can reasonably be taught using asynchronous techniques. This would enable "just-in-time" training for soldiers who need their skills updated, either because of a job change or because of a new item being fielded. This capability would, of course, need to be backed up by instructors available by telephone or e-mail to provide additional assistance. In those cases where hands-on training is still required, mobile teams, locally certified instructors, or supplemental residen-

---

[1]We do not envision hotlines and mobile instructors disappearing, but DL should be able to reduce the need for these more expensive approaches. We also believe that Web-based look-up resources, because of their more universal accessibility and the relatively lower cost of keeping them up to date, will largely replace technical libraries. TRADOC's digital library, which offers manuals, publications, graphic aids, course materials, and some full course materials, receives in excess of 2.5 million hits per week.

tial training will be needed, but DL can clearly reduce these requirements; it can also better prepare students to engage in hands-on training, making that training more efficient when it is delivered.

Along the same lines, just-in-time training can facilitate the acquisition of supplemental skills needed by some selected leaders, both officers and NCOs. We have in mind here the skills required to perform what the military services commonly call "additional duties,"[2] including safety, physical security, environmental protection, load and movement planning and execution, information systems security, and (even) areas such as tax and voting assistance. Many of these supplemental skills can be largely taught using asynchronous DL (or even completely taught, e.g., tax assistance and information systems security[3]). Most Army installations have a locally available course for many of these additional duties; DL could supplement or largely supplant those courses, providing the additional advantages of standardization and ease of scheduling. The latter would be a boon to the Army's operational units. In many cases, training events and deployments include a requirement for a unit to have an officer or NCO certified in one or more additional duties (e.g., safety, preparation of air or rail car loads). More flexibility in scheduling—more readily available training—would obviously make it easier for units to replace these people when they rotate.

It should also be possible to employ certain forms of just-in-time training to meet some unit training requirements. For example, units being deployed to a given region need some basic familiarity with the characteristics of the people, geography, climate, and inherent dangers of that region. However, they do not need that familiarity until they are about to deploy there. Just-in-time DL can make that aspect of deployment preparation easier to accomplish.

---

[2]Officers and NCOs assigned these additional duties are not meant to be sophisticated subject-matter experts. Rather, they are expected to understand the basics needed in each supplemental duty. This basic understanding is supposed to be sufficient to enable them to supervise and direct the unit in accomplishing routine tasks relevant to the duty, to inspect and evaluate the unit's preparedness to perform those tasks as required, and to know when, where, and how to request more expert assistance.

[3]The Army's Signal Center currently offers a DL course in information systems security aimed at producing the skills needed for someone assigned this additional duty, typically on a brigade or battalion staff. It is entirely asynchronous, with an instructor available on a help line. It has so far trained upwards of 2,000 people.

The benefits of informal learning at home station also include the possibility that materials and training could be shared informally with other members of the home unit or organization. CD-based job aids have already been used for unit training; Web-based exercises and training sessions (e.g., for staffs or staff subelements training on planning processes) could easily take their place, offering greater interactivity and updating capability. This would also be a way to capitalize on the DL phase of professional development courses. For example, an officer or NCO slated for attendance at an advanced course could use some of the DL materials to present a class or run a short seminar for members of the unit. This individual would thereby get practice in presenting training, learn the material himself (thus completing part of the DL requirement), and simultaneously contribute to unit training.

GENERAL EDUCATION BENEFITS

We have so far focused on the potential for DL to enable more efficient acquisition of needed military and technical skills. A less obvious and largely untapped potential is for DL to help the military services raise their general levels of education. The more complex and diverse missions the military services face today put a greater premium on general education, requiring more than ever that leaders have a broad perspective, a fuller understanding of the world environment and its historical context, and knowledge of civilian institutions. The growing complexity of potential missions and of the technology employed to accomplish them increases as well the need for well-developed decisionmaking and critical thinking skills. These skills are developed and enhanced through education. But at the same time that requirements appear to be growing for a more highly educated officer and enlisted force, resource constraints and conflicting time demands are working in the opposite direction.

Traditionally, the military services have provided only professional military education for their enlisted personnel; civil schooling has been expected to occur as part of professional self-development on the service member's own time. Today's operational pace and the accompanying greater frequency of deployments make it more and more difficult for service members to find the time (or, more to the point, a sufficiently long uninterrupted span of time) to take advantage of traditional residential instruction offered at universities, colleges, and local institutions. While this is particularly true for those in operational field units, it holds for those in institutional support organizations as well.

The situation is similar for officers, although in the case of officers we are dealing with more advanced degrees, mostly masters and a few at the doctorate level. Historically, officers enter their service with a bachelor's degree and some basic military education; subsequently, they receive professional military education at discrete intervals, and some are selected for full-time funded attendance at civilian academic institutions to study for advanced degrees. The rest attain advanced degrees on their own, if at all. This approach may not be adequate to the demands of today and tomorrow. The military's need for officers with advanced education is increasing, and—as with the enlisted force—the operational pace and frequent deployments make it more difficult to find time to take courses leading to an advanced degree or to the technical certification needed in some career fields.

DL has significant potential to help the military services overcome the challenges summarized above. It can deliver education in smaller packages and provide access to educational materials for students at dispersed military bases and deployed locations. Students can use these packages when and where they can find the time to take advantage of them; thus, the packages are innately easier to schedule. Also, by reducing the importance of geographical separation in selecting educational institutions, DL will serve to make the entire process of offering education more competitive for the offerors.[4] This, in turn, will increase the leverage available to the military services and their members as customers, making it possible for them to receive higher-quality education at the same or lower costs.

While we do not believe DL can fully supplant residential instruction in civilian education any more than it can in military training, it can significantly reduce the need to send service members to blocks of residential instruction and, thus, make the overall process of raising education levels easier. Much remains to be learned about the costs

---

[4]This is an observation with implications that go well beyond the education of military service members.

and benefits of different ways to capitalize on DL in this regard. Answering the numerous questions that have arisen and are still to surface will require an extensive effort by the Department of Defense, the military services, and the civilian academic community. The result of this effort can well be a cost-effective program for improving general levels of education throughout the military services, despite the challenges of resource constraints, operational pace, and deployments.

Chapter Five

CONCLUSIONS AND POLICY IMPLICATIONS

In this chapter we present the conclusions of our analyses of the potential of distance learning to enhance Army personnel readiness, and we discuss policy and implementation issues that must be resolved for distance learning to help achieve personnel readiness payoffs.

DL CAN HELP REDUCE PERSONNEL SHORTAGES AND ENHANCE STABILITY

Based on our first set of analyses, we conclude that the Army can capitalize on distance learning to help reduce personnel shortages and improve fill rates in military occupational specialties, as part of existing strategies to accomplish these objectives. DL is well suited to facilitate reclassification training, support the cross-training of soldiers and the consolidation of MOSs, and accelerate professional education offered to NCOs. DL can help make these existing strategies easier to implement—thus more attractive to soldiers and commanders—and more cost-effective for the Army.

DL can be used to help fill personnel gaps at both SL1 and NCO levels. Moreover, the strategies can improve fill rates without increasing force size, which is not the case with more broadly based accession or retention strategies. Finally, DL-based strategies can decrease the per-soldier cost of reducing shortages, both in terms of the actual costs of training a soldier (by approximately $4,500 in training cost per reclassification) and in terms of later costs of SRBs. While the savings are not budget savings in most instances, DL can increase the efficiency of the overall process of reducing MOS shortages and can lead to the avoidance of a significant amount of the future cost of the Army's efforts to reduce personnel shortages.

From our second set of analyses, we conclude that converting portions of the Army's resident professional development courses to DL can improve stability and reduce turbulence among officers and NCOs and thereby make them more available to their units. In most cases we considered, this applies to their families as well. The estimated increase in man-years available for unit duties is more than 900 working man-years,[1] even after allowing time to participate in DL courses at home station.

Our analysis points to retaining current patterns for officer career courses, with appropriate shortening of the residential advanced course phase as DL conversions make that possible. Potential PCS savings would likely be more than offset by increased TDY costs if these courses were shifted to a TDY mode, and the family impact of increased separations would be significant. Courses that are already conducted in a TDY mode also show potential for decreasing the time soldiers spend away from home and from unit duties, with modest but unambiguous savings potential as well.

REALIZING THIS POTENTIAL REQUIRES CHANGES IN POLICY EMPHASIS

While DL does offer a great deal of promise to help the Army deal with some of its personnel readiness concerns, realizing that promise requires careful planning and implementation of DL programs. Enhancing officer and NCO professional stability means making stability enhancement a major objective of DL and emphasizing this application in the active force. To aid in strategies for reducing personnel shortages, the Army must make these applications a policy priority and select courses to convert that will help the most to reduce personnel shortages. Also, as "critical MOSs" are subject to change, the Army must be prepared to shift priorities and develop new courseware in a timely manner. As of FY99, only 19 of 51 MOSs deemed "critical" with respect to personnel fill had courses in the

[1]Combining our minimum estimates for officer advanced courses (115 man-years) and other courses (800 man-years).

pipeline for conversion to DL. Also, of 44 MOSs that we believe are prime consolidation candidates, only 6 have courses slated for conversion to DL before FY03.

The DL program should make the fullest possible use of emerging learning technologies to help reduce learning time (and, thus, shorten courses), and to allow the completion of significant portions of the training at home station. Furthermore, using DL in this way means creating DL courses that are attractive to students, commanders, and the Army and have sufficient flexibility to easily integrate into varying soldier career paths. This calls for relatively more emphasis on Web-based asynchronous modules wherever possible—and careful scrutiny of those instances in which VTT or other synchronous modes of delivery appear to be required. In particular, for cases in which VTT use can be reduced, this can also reduce associated costs and make facilities more available for other video communication uses. In addition, the DL program should strive to provide sufficient student support to ensure timely course completion, and sufficient administrative support for scheduling, monitoring, and recording training results. Failures in these areas have plagued past DL programs in industry and academia.

We also observe that care must be exercised in selecting course segments to be taught using DL: much of the training the Army needs to conduct is simply not amenable to this approach. For those skills that can be imparted using DL, planners must realistically assess the amount of time needed to train these skills, and the Army must ensure that soldiers have this time made available to them. This means providing for dedicated ("fenced") time and may require promulgation of both Army-wide and local policies describing the requirement to make such provisions. Also, course scheduling will remain an item of interest for students, training managers, and commanders; if anything, DL programs make close coordination and timely use of the Army Training Requirements and Resources System (ATRRS) even more important.

For those applications deemed suitable for DL, further care is needed to ensure that the quality and effectiveness of training is maintained. DL's ability to deliver the training the Army needs—with no diminution in quality—is a major premise underlying our analyses and, for that matter, the entire DL program. As we have noted, past research supports the contention that DL can provide training as effectively as the classroom training it replaces, and possibly more efficiently in some cases.[2] Generally, studies of various forms of DL have pointed toward a tradeoff between superior performance and reduced training time, compared with the RL courses they are designed to replace. A likely reason for this effect is that the DL instruction is self-paced, so students only spend as much time as they need to achieve a given performance standard. The wider implication of this effect is that training managers and the Army leadership will face a large number of choices with regard to the tradeoff between reduced training time and improved training effectiveness. These choices will be complicated by the desire in some cases to capture the values of group discussion and close interactions with expert instructors, none of which can be provided quite as well with DL as they can with RL. Thus, we emphasize again the need for continued care in selecting course segments for conversion to DL. Also, as with any RL curriculum, periodic refinement and adjustment of DL course content will also be necessary to maintain currency.

It will also be important for the Army to avoid premature confidence in any major savings estimates. In particular, we caution against planning on large PCS savings. DL will not reduce PCS moves unless an independent decision is made to convert courses from PCS to TDY. Even in those cases where moves might be reduced, any savings will be largely or completely offset by increased TDY costs. Furthermore, converting long courses to TDY mode will add more family separations. Finally, estimates of eventual savings are based on DL conversions that have not taken place and will not occur until the Army has worked through its conversion schedule.

The Army will find it fruitful to undertake a more detailed study of the costs associated with institutional training, including some of the factors used here, but also extending to school resources: instructors, support, training materials, courseware development and maintenance, and longer-term facilities and other capital costs. Our previous DL research has found these costs—even when considering only the portions of them associated with DL—to be considerable.

---

[2]See, for example, Keene and Cary (1992) and Orlansky and String (1979, 1981).

Finally, we note that the tasking to achieve the full potential of DL as part of the Army's ongoing transformation—either to reduce personnel shortages or to improve stability ("determine means to attain and maintain required levels of personnel fill and stabilization throughout [the] transformation")—belongs to the DCSPER. As such, ODCSPER must take a more proactive role in ensuring that the current DL program is refocused to fully address the human resource needs of the Army.

REFERENCES

Farris, Hilary, William L. Spencer, John D. Winkler, and James P. Kahan. *Computer-Based Training of Cannon Fire Direction Specialists.* Santa Monica, CA: RAND, MR-120-A, 1993.

Hix, William M., Herbert J. Shukiar, Janet M. Hanley, Richard J. Kaplan, Jennifer H. Kawata, Grant N. Marshall, and Peter J.E. Stan. *Personnel Turbulence: The Policy Determinants of Permanent Change of Station Moves.* Santa Monica, CA: RAND, MR-938-A, 1998.

Keene, S. Delane, and James S. Cary. "Effectiveness of Distance Education Approach to U.S. Army Reserve Component Training," in Michael G. Moore (ed.), *Distance Education for Corporate and Military Training.* University Park, PA: Pennsylvania State University, 1992.

Leonard, Henry A., John D. Winkler, Anders Hove, Emile Ettedgui, Michael G. Shanley, and Jerry Sollinger. *Enhancing Stability and Professional Development Using Distance Learning.* Santa Monica, CA: RAND, MR-1317-A, 2001.

Orlansky, Jesse. "Effectiveness of CAI: A Different Finding." *Electronic Learning,* Vol. 3, No. 1, September 1983.

Orlansky, Jesse, and Joseph String. *Cost-Effectiveness of Computer-Based Instruction in Military Training.* Alexandria, VA: Institute for Defense Analyses, IDA Paper P-1375, April 1979.

Orlansky, Jesse, and Joseph String. "Computer-Based Instruction for Military Training." *Defense Management Journal,* Second Quarter 1981, pp. 46–54.

Phelps, Ruth H., Rosalie A. Wells, Robert L. Ashworth, Jr., and Heidi A. Hahn, "Effectiveness and Costs of Distance Education Using Computer-Mediated Communication," in Michael G. Moore (ed.), *Distance Education for Corporate and Military Training.* University Park, PA: Pennsylvania State University, 1992.

Program Management Office, TADLP. *The Army Distance Learning Program Economic Analysis,* September 13, 2000.

Shanley, Michael G., Henry A. Leonard, and John D. Winkler. *Army Distance Learning: Potential for Reducing Shortages in Enlisted Occupations.* Santa Monica, CA: RAND, MR-1318-A, 2001.

Shukiar, Herbert J., John D. Winkler, and John E. Peters. *Enhancing the Retention of Army Noncommissioned Officers.* Santa Monica, CA: RAND, MR-1186-A, 2001.

Wild, William G., Jr., and Bruce R. Orvis, *Design of Field-Based Crosstraining Programs and Implications for Readiness.* Santa Monica, CA: RAND, R-4242-A, 1993.

Winkler, John D., Herbert J. Shukiar, James A. Dewar, Matthew W. Lewis, Beth Benjamin, Jerry M. Sollinger, John E. Peters, and Harry J. Thie. *Future Leader Development of Army Noncommissioned Officers: Workshop Results.* Santa Monica, CA: RAND, CF-138-A, 1998.